Math Mammoth
Grade 7 Tests and
Cumulative Revisions

for the complete curriculum
(International Series)

Includes consumable student copies of:

- Chapter Tests
- End-of-year Test
- Cumulative Revisions

By Maria Miller

Contents

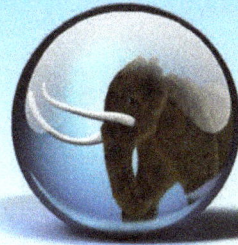

Grade 7, Chapter 1

End-of-Chapter Test

Instructions to the student:

Do **not** use a calculator. Answer each question in the space provided.

Instructions to the teacher:

You can give partial credit for partial solutions. The total is 21 points, so divide the student's score by 21 and multiply by 100 to get a percent score. For example, if the student scores 17, divide $17 \div 21$ to get 0.8095. The percent score is 81%.

Question	Max. points	Student score
1	2 points	
2	2 points	
3	1 point	
4	2 points	
5	2 points	

Question	Max. points	Student score
6	3 points	
7	4 points	
8	2 points	
9	3 points	
TOTAL	21 points	/ 21

Chapter 1 Test

1. Write an expression with three terms. The coefficient of the first term is 2 and of the second term is 5. The last term is the constant 9. The variable part of the first term is s squared, and the variable of the second term, t.

2. Evaluate the expressions.

a. $2(7 - x)^2$, when $x = 2$	**b.** $\dfrac{1}{g} + \dfrac{g + 1}{3}$, when $g = 6$

3. Name the property of arithmetic illustrated by the fact that $5(z + 4)$ is equal to $5z + 20$.

4. Draw a diagram of two rectangles to illustrate that the product $5(z + 4)$ is equal to $5z + 20$.

5. Write each expression as a product (factor it).

a. $7x + 21 = $ _____ $(x + $ _____ $)$	**b.** $24k + 80 = $

6. Simplify the expressions.

a. $v + 5 + v + v + v$	**b.** $v \cdot 5 \cdot v \cdot v \cdot v$	**c.** $8x + 5 - 3x - 2$

7. Write an equation and solve it using guess and check.

    a. Seven times the quantity x minus one equals 14.

    b. Two less than x squared equals 23.

8. Write an expression for each situation.

    a. Abigail bought x bags of nuts for $3 a bag. She paid with a $50 banknote. What was her change?

    b. A pair of jeans that costs p dollars is discounted by 1/10 of its price. What is the discounted price?

9. **a.** Write and simplify an expression for the total area of the shape.

    b. Evaluate your expression when $x = 2$ cm.

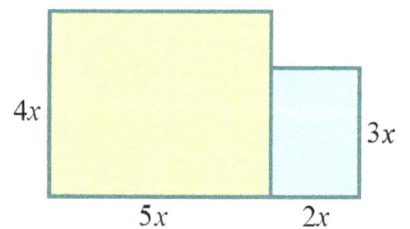

4x   3x   5x   2x

    c. Write an expression for the perimeter of the shape.

Grade 7, Chapter 2

End-of-Chapter Test

Instructions to the student:

Do **not** use a calculator. Answer each question in the space provided.

Instructions to the teacher:

You can give partial credit for partial solutions. The total is 33 points, so divide the student's score by 33 and multiply by 100 to get a percent score. For example, if the student scores 28, divide 28 ÷ 33 to get 0.8484. The percent score is then 85%.

Question	Max. points	Student score
1	2 points	
2	1 point	
3	4 points	
4	5 points	
5	2 points	
6	2 points	
7	1 point	
8	1 point	

Question	Max. points	Student score
9	1 point	
10	3 points	
11	4 points	
12a	1 point	
12b	1 point	
12c	2 points	
13	3 points	
TOTAL	33 points	/ 33

Chapter 2 Test

1. Illustrate the sum $-5 + 7$ in two ways: (1) using counters, and (2) using a number line.

2. Write a sum where you add a number and its opposite.

3. Add or subtract.

a. $11 + (-8) =$	**b.** $11 - (-8) =$	**c.** $-11 - 8 =$	**d.** $-11 + 8 =$

4. Add or subtract.

a. $61 + (-82) =$	**b.** $-55 - (-29) =$	**c.** $43 - 189 =$
d. $-15 + (-9) + 3 + (-8) =$		**e.** $2 - (-11) + (-13) + (-7) + 10 =$

5. Explain a real-life situation for the calculation -80 m $- 30$ m $= -110$ m.

6. Write an addition or subtraction, and solve.

   Aiden owed $21 to his Dad. Then he paid $15. Then he borrowed another $35.
   Then he made another payment of $50. What is his balance now?

7. Allison's mum designed a point system for Allison where she would get positives for doing her chores and school work well and negatives for doing her chores and school work poorly.

   On Tuesday, Allison "earned" five negative points. On Monday, her final tally had been 6 points. How many points better did Allison do on Monday than on Tuesday?

8. The distance between x and -5 is nine units.
   Find two possible values for x.

9. Give an example of a night-time and a daytime temperature where both are negative and have a difference of 9 degrees.

10. Fill in the missing numbers.

a. $-7 \cdot$ _____ $= -35$	**b.** $2 \cdot$ _____ $= -100$	**c.** $-50 \div$ _____ $= -10$

11. Simplify. Give the answer in lowest terms and as a mixed number if applicable.

a. $15 \div (-3) =$	**b.** $-2 \div (-10) =$
c. $-20 \div 6 =$	**d.** $72 \div (-48) =$

12. Find the value of each expression.

a. $-3 + \dfrac{15}{(-5)}$	**b.** $-2 \cdot \dfrac{(-24)}{4}$	**c.** $-100 - \dfrac{36}{(-9)} - (-20)$

13. Find the value of the expressions when $x = -3$ and $y = 4$.

a. $xy - 2$	**b.** $x^2 + 1$	**c.** $-2(y - 5)$

Grade 7, Chapter 3

End-of-Chapter Test

Instructions to the student:

Use of a calculator with basic functions ***is*** allowed. Answer each question in the space provided.

Instructions to the teacher:

You can give partial credit for partial solutions. The total is 19 points, so divide the student's score by 19 and multiply by 100 to get a percent score. For example, if the student scores 14, divide 14 ÷ 19 to get 0.7368. The percent score is 74%.

Question	Max. points	Student score
1	8 points	
2	equation: 1 point	
	solution: 1 point	
3	equation: 1 point	
	solution: 1 point	

Question	Max. points	Student score
4	3 points	
5	2 points	
6	2 points	
TOTAL	19 points	/ 19

Chapter 3 Test

1. Solve. Check your solutions.

a. $x + 8 = -13$	**b.** $4 - (-2) = -y$
c. $18 - x = -1$	**d.** $2 - 6 = -z + 5$
e. $5x + 3x = -31 + (-9)$	**f.** $7 \cdot (-2) = 3y - 15y$
g. $\dfrac{x}{10} = -17 + 5$	**h.** $-13 = \dfrac{c}{-7}$

Write an equation for the problem. Then solve it. Don't write just the answer.

2. Four kilograms of chicken breast cost $38.00. How much does one kilogram cost?

Write an equation for the problem. Then solve it. Don't write just the answer.

3. Noah's suitcase is 4.6 kg heavier than Bill's. If Noah's suitcase weighs 28.7 kg, then how much does Bill's weigh?

4. Use the formula $d = vt$ to solve the problem.

A ferry travels at a constant speed of 18 km/h. How long will it take to cross a river, a distance of 600 *metres*?	d = v t ↓ ↓ ↓

5. How far can you travel in 1 hour 25 minutes, bicycling at a constant speed of 15 km/h?

6. Find the average speed of an aeroplane that flies 4455 kilometres in 4 1/2 hours.

Grade 7, Chapter 4

End-of-Chapter Test

Instructions to the student:

Do **not** use a calculator. Answer each question in the space provided.

Instructions to the teacher:

You can give partial credit for partial solutions. The total is 51 points, so divide the student's score by 51 and multiply by 100 to get a percent score. For example, if the student scores 37, divide $37 \div 51$ to get 0.7255. The percent score is 73%.

Question	Max. points	Student score
1	5 points	
2	2 points	
3	4 points	
4	2 points	
5	2 points	
6	4 points	
7	6 points	
8	8 points	

Question	Max. points	Student score
9	2 points	
10	3 points	
11	3 points	
12a	2 points	
12b	2 points	
12c	3 points	
12d	3 points	
TOTAL	51 points	/ 51

Chapter 4 Test

1. Mark these numbers on the number line: $-\dfrac{2}{5}, \quad -1\dfrac{4}{5}, \quad -\dfrac{11}{5}, \quad -\dfrac{1}{10}, \quad -\dfrac{19}{10}$.

2. Write each decimal as a fraction.

   a. 5.001

   b. −2.0482

3. Write each fraction as a decimal.

   a. $-\dfrac{8}{5}$

   b. $-\dfrac{47}{10\,000}$

   c. $\dfrac{787}{10}$

   d. $-\dfrac{5624}{100}$

4. Which is more, $0.\overline{6}$ or 0.6?
   How much more?

5. Find 15% of 3/4.

6. Write as decimals, using a line over the repeating part (if any). Use long division.

a. $\dfrac{7}{6}$	**b.** $\dfrac{5}{36}$

7. Calculate.

a. $-1.26 - (-3.45)$	**b.** $1.8 - 3.25$	**c.** $-0.42 + 10.7 + (-9.8)$
d. $-0.06 \cdot 0.05$	**e.** $(-0.5)^3$	**f.** $1.2 \cdot (-0.4) - 3 \cdot (-0.8)$

8. Find the value of each expression. Give it as a mixed number if applicable.

a. $\dfrac{5}{9} + \left(-\dfrac{2}{3}\right)$	**b.** $-\dfrac{1}{10} - \dfrac{6}{9}$
c. $\dfrac{\frac{2}{5}}{-4}$	**d.** $\dfrac{\frac{9}{10}}{\frac{1}{6}}$

9. Two-thirds of a number is −5.66.
   What is the number?

10. Give a real-life context for the calculation $\frac{1}{3} \cdot 12.75$. Then solve.

11. Give a real-life context for the calculation $0.8 \cdot (-1530)$. Then solve.

12. Find the value of each expression. If you use fraction arithmetic, give your answer as a mixed number when applicable. *Hint:* You can use estimation to check that your final answer is reasonable.

a. $-6\frac{1}{9} \div \left(-\frac{4}{3}\right)$	**b.** $-0.9 \div 0.011$
c. $-1\frac{7}{10} - 0.5 \cdot \frac{4}{5}$	**d.** $1.1 \cdot \left(-\frac{5}{8} + \frac{1}{2}\right) \cdot (-0.4)$

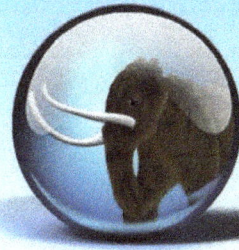

Grade 7, Chapter 5

End-of-Chapter Test

Instructions to the student:

Do **not** use a calculator. Answer each question in the space provided.

Instructions to the teacher:

You can give partial credit for partial solutions. The total is 28 points, so divide the student's score by 28 and multiply by 100 to get a percent score. For example, if the student scores 19, divide 19 ÷ 28 to get 0.6786. The percent score is 68%.

Question	Max. points	Student score
1	12 points	
2	equation: 1 point	
	solution: 2 points	
3	equation: 1 point	
	solution: 2 points	

Question	Max. points	Student score
4	3 points	
5	4 points	
6a	1 point	
6b	1 point	
6c	1 point	
TOTAL	28 points	/ 28

Chapter 5 Test

1. Solve. Give your answers as fractions or whole numbers (not decimals).

a. $\quad -2 \;=\; 6x + 5$	**b.** $\quad 6(x + 2) - 1 \;=\; -9$
c. $\quad \dfrac{3x}{5} \;=\; 24$	**d.** $\quad \dfrac{y}{3} \;-\; 21 \;=\; -5$
e. $\quad 6\left(t - \dfrac{2}{3}\right) \;=\; 47$	**f.** $\quad \dfrac{3}{4}(x - 12) \;=\; -21$

2. Ethan purchased 24 cookies and a loaf of bread for a total of $10.29 (which was rounded to $10.30). He didn't pay attention to the cost of the cookies but he remembered that the bread cost $5.25. Find the cost of one cookie by writing an equation and solving it.

3. Currently, Alice exercises by jogging five times around a 1.15-km loop among the city blocks. She'd like to increase the total distance she jogs to 7 km, yet keep going around a loop five times. How much longer should the individual loop be? Solve the problem by using an equation.

4. A student wrote this equation for the problem on the right.

$$2(x + 9) = 78$$

Is the student correct? If not, correct the equation and solve the problem.

The perimeter of a parallelogram is 78 cm. The base sides are 9 cm longer than the other two sides. Find the side lengths of the parallelogram.

5. Solve each inequality and plot its solution set on the number line.
   Write the appropriate numbers for the tick marks.

a. $\quad 3x + 5 \quad < \quad 68$	**b.** $\quad -10x - 17 \quad \geq \quad 103$

6. Abner is building a shed. The building permit for it limits its height to a maximum of 2.85 metres above the surface of the ground. He poured an above-ground foundation that was 10 cm thick, and the flat roof will add 15 cm to the height of the wall. He is going to build with concrete blocks that are 20 cm high each (including the mortar).

   a. Write an inequality to calculate how many rows of block he can lay without the shed exceeding the maximum height permitted.

   b. Solve the inequality.

   c. Draw a number line and plot the solution set.

Grade 7, Chapter 6

End-of-Chapter Test

Instructions to the student:

A basic calculator is allowed. Answer each question in the space provided.

Instructions to the teacher:

You can give partial credit for partial solutions. The total is 23 points, so divide the student's score by 23 and multiply by 100 to get a percent score. For example, if the student scores 20, divide $20 \div 23$ to get 0.869565. The percent score is 87%.

Question	Max. points	Student score
1	2	
2a	2	
2b	1	
3	3	
4	2	
5	2	
6	3	

Question	Max. points	Student score
7	2	
8a	1	
8b	1	
8c	1	
9a	1	
9b	2	
TOTAL	23	/ 23

Chapter 6 Test

You may use a calculator for all the problems in this test.

1. Chloe bicycled 21 kilometres in 1 1/2 hours.
   Write a rate for her speed and simplify it to find the unit rate.

2. Mason poured 1/3 of an envelope of chocolate drink powder into 2/3 cups of water.

   a. Find the unit rate (as envelope(s) per cup).

   b. What does the unit rate signify?

3. Write a proportion for the following problem and solve it.

   A bag of 52 kg of wheat costs $169.   _____ = _____
   What would 21 kg of wheat cost?

4. Find the value of S to one decimal digit.

$$\frac{4}{S} = \frac{7.9}{12}$$

5. The figures are similar. Find the length of the side labelled with x.

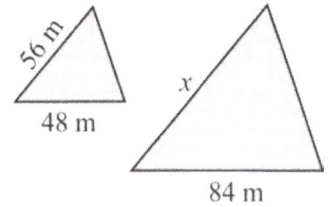

6. The figure on the right is a house plan with a scale of 1 cm = 60 cm. Find the area of the house in reality, in square metres.

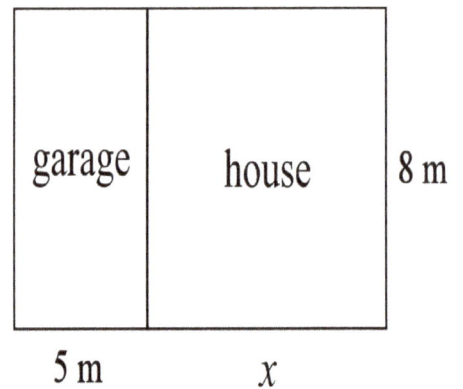

7. A town map has a scale of 1:45 000.

   a. A street in this town is 850 m long. How long is that street on this map?

   b. How long in reality is a road that measures 5.4 cm on the map?

8. The graph on the right depicts the distance that a running fox covers as time passes.

   a. State the unit rate (including the units of measurement) for this situation.

   b. Plot the point that corresponds to the unit rate.

   c. Write an equation for the line.

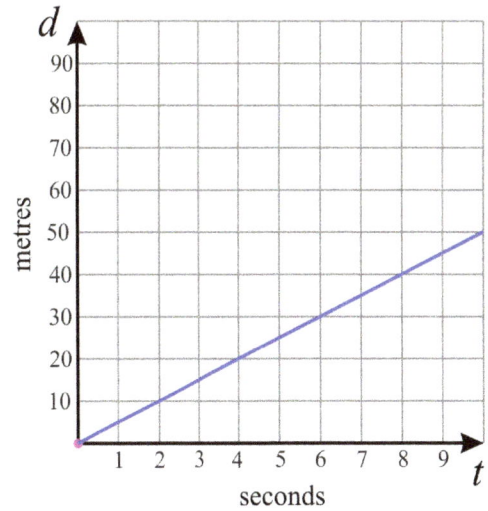

9. The equation $S = 15t$ tells us the salary (S, in dollars) of a worker who works for t hours.

   a. What is the unit rate in this situation?

   b. Graph the equation $S = 15t$ in the grid. Label the axes. Choose the scaling for the two axes so that the point corresponding to working for 10 hours will fit on the grid.

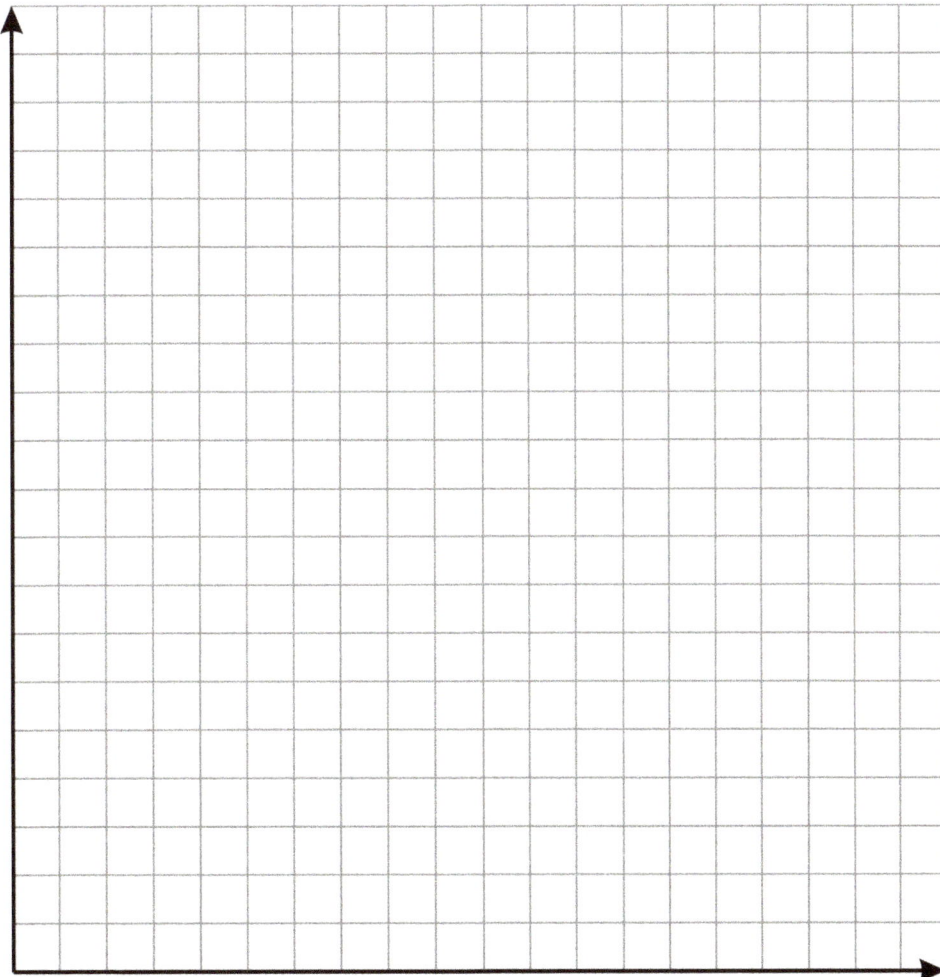

Grade 7, Chapter 7

End-of-Chapter Test

Instructions to the student:

A basic calculator is allowed. Answer each question in the space provided.

Instructions to the teacher:

You can give partial credit for partial solutions. The total is 17 points, so divide the student's score by 17 and multiply by 100 to get a percent score. For example, if the student scores 11, divide $11 \div 17$ to get 0.6471. The percent score is 65%.

Question	Max. points	Student score
1	3	
2	3	
3	1	
4	2	
5	2	

Question	Max. points	Student score
6	2	
7	2	
8	1	
9	1	
TOTAL	17	/ 17

Chapter 7 Test

You may use a basic calculator for all the problems in this test.
If not otherwise specified, give your answers that are percentages to the tenth of a percent.

1. The price of these items is changing. Find the new price or the discount percentage.

a. Price: $110 12% discount New price: $_____	**b.** Price: $5000 2.4% increase New price: $_____	**c.** Price: $90 _____ % discount New price: $59

2. The Jefferson family bought three children's tickets and two adult's tickets to the county fair. They got a 5% discount on the total purchase price before tax. Lastly, a 6.2% sales tax was added to the total. If the normal price of a child's ticket is $10 and an adult's ticket is $20, find the cost of tickets for the family.

3. This year the college has 1210 students—an increase of 6.6% from last year. How many students did the college have last year?

4. Mary's dog weighed 25 kg, but then it got sick and lost 2.3 kg.

   a. What percentage of body weight did the dog lose?

   b. Mary weighs 58 kg. If Mary lost the same percentage of her body weight as what the dog did, how much would Mary weigh?

5. A rectangular playground area measures 5 m by 6.5 m. It is enlarged so that it becomes 7.2 m by 10 m. What is the percentage of increase in its area?

6. Jeff bought five packages of tile at 22% off. His total bill was $152.10. What was the price of one package of tile *before* the discount?

7. A merchant is currently selling dehumidifiers for $145.50. He wants to increase the price by some amount, so that when he offers the dehumidifier for 25% off, the customer will pay $125 for it. By how much should he increase the price?

8. Jacqueline deposited $2500 into a savings account that pays a yearly interest rate of 4.4%. Calculate how much her account will contain after three years.

9. Michael borrowed $35 000 for ten years. At the end of those years he paid the bank back $65 800. What was the interest rate?

Grade 7, Chapter 8

End-of-Chapter Test

Instructions to the student:

Do **not** use a calculator for problems 1-6. You **may** use a basic calculator for problems 7-13. Answer each question in the space provided.

Instructions to the teacher:

You can give partial credit for partial solutions. The total is 26 points, so divide the student's score by 26 and multiply by 100 to get a percent score. For example, if the student scores 21, divide 21 ÷ 26 to get 0.80769. The percent score is 81%.

Question	Max. points	Student score
1	1	
2	2	
3a	2	
3b	1	
4	3	
5	2	
6	2	

Question	Max. points	Student score
7a	2	
7b	1	
8	3	
9	1	
10	1	
11	2	
12	3	
TOTAL	26 points	/ 26

Chapter 8 Test

1. Draw two angles that are complementary.

2. **a.** Find the measure of angle EOF.

   b. Find the measure of angle FOG.

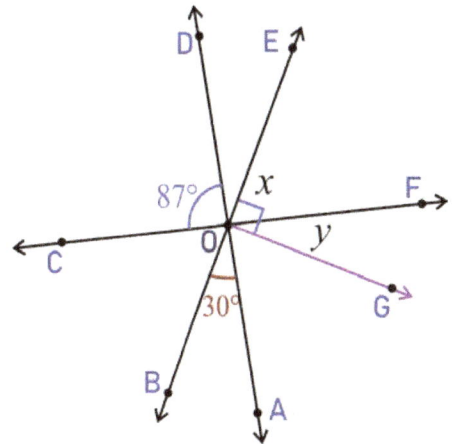

3. **a.** Find the value of x.

   b. Find the measure of angle COD.

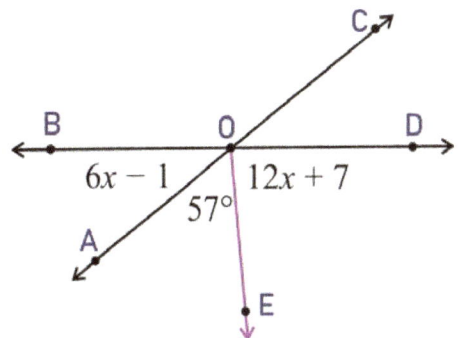

4. A triangle has an 80° angle and two 11-cm sides.

   Does this information determine a unique triangle? If so, draw it.

   If not, draw two non-congruent (differently shaped) triangles that satisfy the conditions.

5. Draw any equilateral triangle using only a compass and a straightedge.
   You can choose how long the sides are.

6. A large circular wall clock has a diameter of 40.0 cm. Find its area to the nearest ten square centimetres.

7. One hectare is 10 000 square metres. Use that information and your knowledge of geometry to calculate the area of this trapezoidal plot:

   a. in square metres

   b. in hectares, to the hundredth of a hectare.

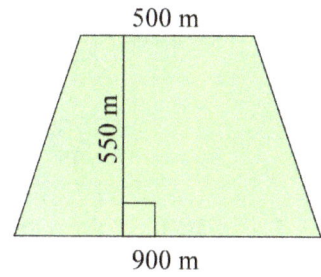

500 m

550 m

900 m

8. This shape consists of a regular hexagon and six semi-circles. Find its area to the nearest ten square centimetres.

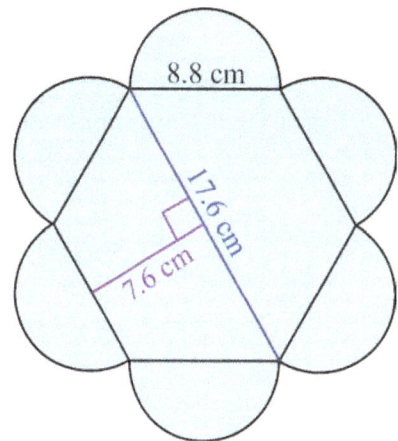

8.8 cm

17.6 cm

7.6 cm

9. A rectangular prism is cut with a plane that is perpendicular to the prism's base. What figure is formed at the cross-section?

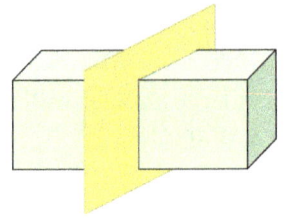

10. A rectangular pyramid is cut with a plane that is parallel to the pyramid's base. What figure is formed at the cross-section?

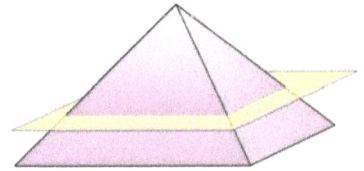

11. Calculate the volume of this water tank in <u>cubic metres</u>.

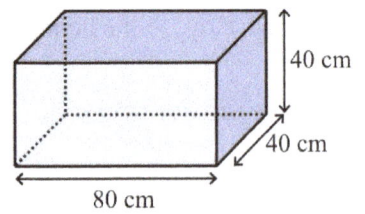

40 cm

40 cm

80 cm

12. Calculate the surface area of this cylindrical bucket for toys to the nearest ten square centimetres.

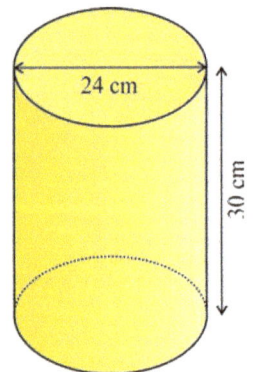

24 cm

30 cm

math MAMMOTH

Grade 7, Chapter 9

End-of-Chapter Test

Instructions to the student:

A basic calculator is allowed. Answer each question in the space provided.

Instructions to the teacher:

You can give partial credit for partial solutions. The total is 22 points, so divide the student's score by 22 and multiply by 100 to get a percent score. For example, if the student scores 16, divide $16 \div 22$ to get 0.727272.... The percent score is 73%.

Question	Max. points	Student score
1	4	
2a	3	
2b	1	
2c	1	
2d	1	
3	3	

Question	Max. points	Student score
4a	1	
4b	3	
4c	1	
5	2	
6a	1	
6b	1	
TOTAL	22	/ 22

Chapter 9 Test

You may use a calculator for all the problems in this test.

1. You roll a number cube with numbers 1, 2, 3, 4, 5, and 6 printed on the faces.
   Find the probabilities as fractions.

    a. P(not 5)

    b. P(2 or 6)

    c. P(less than 9)

    d. P(not 2 nor 5)

2. Two spinners are spun.

    a. In the space below, draw a tree diagram showing all the
    possible outcomes of this experiment.

    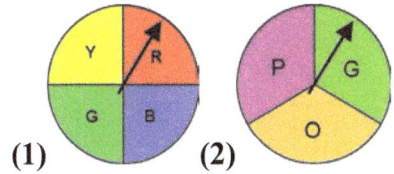

    (1)   (2)

    Then find the probabilities.

    b. P(yellow; purple)

    c. P(red or yellow; orange)

    d. P(not red; not orange)

3. Two dice are rolled. Find the probabilities of these events:

    a. You get a sum of six on the two dice.

    b. You get less than 3 on each dice.

    c. One dice is 6 and the other is not (in either order).

4. Logan and Alex tossed two coins 400 times.

    a. List all the possible outcomes when two coins are tossed just one time.

    b. Here are Logan's and Alex's results. Calculate and fill in the table the experimental and theoretical probabilities to the nearest tenth of a percent.

	Frequency	Experimental probability	Theoretical probability
TT	5		
TH	8		
HT	182		
HH	205		
TOTALS	400		

    c. Suggest a reason for the large discrepancy between the experimental and theoretical probabilities.

5. What is the probability of getting tails, tails, tails when you toss a coin three times in a row?

6. Lily and Grace placed some stuffed animals in a bag. Then they randomly pulled out one animal and put it back, and repeated this 120 times. Here are their results:

Animal	Frequency
Elephant	58
Giraffe	29
Bear	17
Cat	11
Bird	5
Totals	**120**

a. Based on their results, what is the approximate probability of pulling the cat out of the bag?

b. If this experiment was repeated 300 times, approximately how many times should they expect to get the bear?

Grade 7, Chapter 10

End-of-Chapter Test

Instructions to the student:

Do **not** use a calculator. Answer each question in the space provided.

Instructions to the teacher:

You can give partial credit for partial solutions. The total is 20 points, so divide the student's score by 20 and multiply by 100 to get a percent score. For example, if the student scores 15, divide $15 \div 20$ to get 0.75. The percent score is 75%.

Question	Max. points	Student score
1a	1	
1b	1	
1c	1	
1d	3	
2	6	

Question	Max. points	Student score
3	4	
4a	1	
4b	3	
TOTAL	20	/ 20

Chapter 10 Test

1. Researchers compared two different methods for losing weight by assigning 50 overweight people to use each method. The side-by-side boxplots show how many kilograms people in each group lost. Note that this situation has to do with entire populations, not samples. The two research groups *are* the populations.

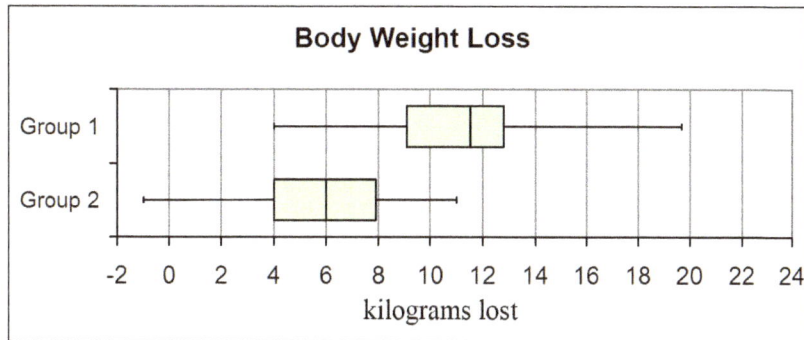

Body Weight Loss

a. Just looking at the two distributions, which group, if any, appears to have lost more weight?

b. Which group, if any, appears to have a greater variability in the amount of weight lost?

c. In group 2, there is one person whose weight loss was −1 kg. What does that mean?

d. Is one of the weight loss methods significantly better than the other?

If so, which one?

Justify your reasoning.

2. Erica is studying the favourite hobbies of adults who live in a small town. She needs a sample of 120 people for her study. Below are listed four possible sampling methods that she could use. Three of the four methods would likely produce a biased sample. Explain which ones they are and why each method is biased.

Sampling method	Biased or not?
(1) Erica generates 120 random numbers between 1 and 1000, and chooses the corresponding people that she meets on the main street of the town. If method (1) is biased, explain why:	
(2) Erica chooses randomly 120 people from a list of the town's residents. If method (2) is biased, explain why:	
(3) Erica places a box and her survey papers in the town's library, and anyone who wants to can fill in the survey questionnaire. If method (3) is biased, explain why:	
(4) Erica chooses randomly 120 people from a list of the town's taxpayers. If method (4) is biased, explain why:	

54

3. An ice cream shop surveyed its customers to find out which new flavours of ice cream to add to their selection. They obtained two samples using their customer database. Here are the results:

	Pineapple	Cappuccino	Peanut Butter	Kiwi	Blueberry	Totals
Sample 1	12	32	15	8	13	80
Sample 2	15	36	10	4	15	80

What can you infer from the results?

4. The two histograms show the age distributions of two groups of people. Below you find the means and the mean absolute deviations for both groups.

a. How much do the means differ?

b. Is the difference in the mean ages significant?

Justify your reasoning.

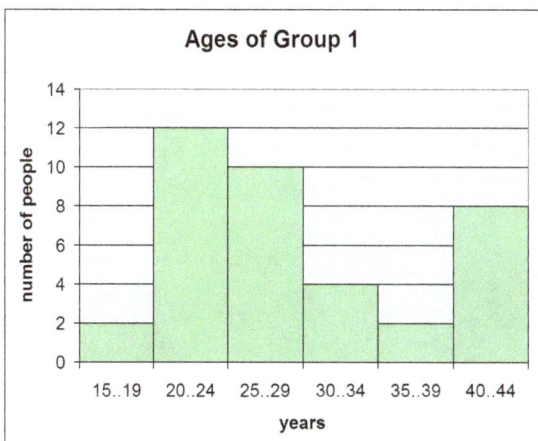

mean 28.6 years   MAD 6.78 years

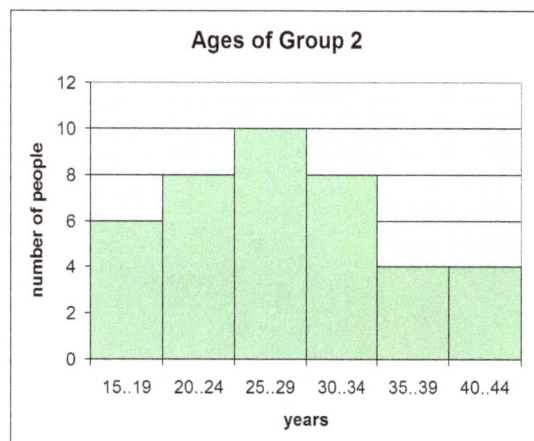

mean 27.55 years   MAD 6.205 years

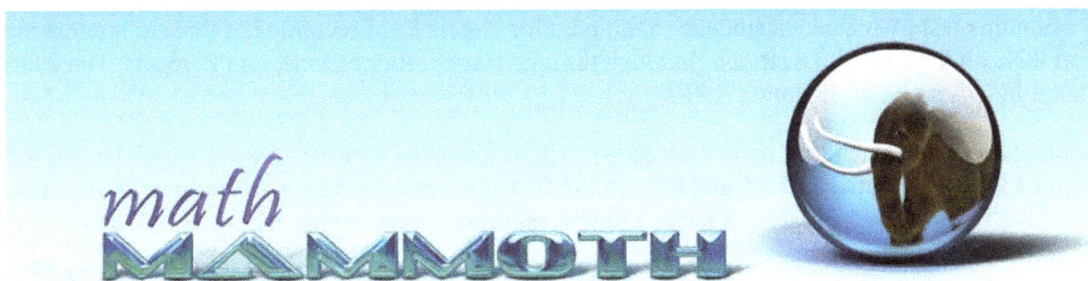

Grade 7 (Pre-algebra) End-of-Year Test
International Version

This test is quite long, because it contains lots of questions on all of the major topics covered in the *Math Mammoth Grade 7* curriculum. Its main purpose is to be a diagnostic test—to find out what the student knows and does not know about these topics.

You can use this test to evaluate a student's readiness for continuing to Math Mammoth Grade 8, or continuing to an Algebra 1 course. In the latter case, it is sufficient to administer the *first four sections* (Integers through Ratios, Proportions, and Percent), because the topics covered in those are prerequisites for algebra or directly related to algebra. The sections on geometry, statistics, and probability are not essential for a student to be able to continue to Algebra 1.

Since the test is so long, I recommend that you break it into several parts and administer them on consecutive days, or perhaps on morning/evening/morning/evening. Use your judgement.

A calculator is *not* allowed for the first three sections of the test: Integers, Rational Numbers, and Algebra.

A <u>basic</u> calculator *is* allowed for the last four sections of the test: Ratios, Proportions, and Percent; Geometry, Probability, and Statistics.

The test is evaluating the student's ability in the following content areas:

- operations with integers
- multiplication and division of decimals and fractions, including with negative decimals and fractions
- converting fractions to decimals and vice versa
- simplifying expressions
- solving certain types of linear equations and linear inequalities
- writing simple equations and inequalities for word problems
- proportional relationships
- unit rates that involve fractions
- basic percent problems, including percentage of change
- working with scale drawings
- drawing triangles
- the area and circumference of a circle
- some angle relationships
- cross-sections formed when a plane cuts a solid
- solving problems involving area, surface area, and volume
- simple probability
- listing all possible outcomes for a compound event
- experimental probability, including designing a simulation
- biased vs. unbiased sampling methods
- making predictions based on samples
- comparing two populations and determining whether the difference in their medians is significant

If you are using this test to evaluate a student's readiness for Algebra 1, I recommend that the student score a minimum of 80% on the first four sections (Integers through Ratios, Proportions, and Percent). The subtotal for those is 120 points. A score of 96 points is 80%.

I also recommend that the teacher or parent revise with the student any content areas in which the student may be weak. Students scoring between 70% and 80% in the first four sections may also continue to Algebra 1, depending on the types of errors (careless errors or not remembering something, versus a lack of understanding). Use your judgment.

You can use the last four sections to evaluate the student's mastery of topics in Math Mammoth Grade 7 Curriculum. However, mastery of those sections is not essential for a student's success in an Algebra 1 course.

The two geometry problems marked with an asterisk (*) are beyond the Common Core Standards for 7th grade.

My suggestion for points per item is as follows.

Question #	Max. points	Student score
Integers		
1	2 points	
2	2 points	
3	2 points	
4a-f (1 pt each)	6 points	
4g-i (2 pts each)	6 points	
5	2 points	
6	2 points	
7	3 points	
subtotal	/ 25	
Rational Numbers		
8	3 points	
9	3 points	
10	4 points	
11	4 points	
12	6 points	
13	2 points	
14	4 points	
subtotal	/ 26	
Algebra		
15	6 points	
16	3 points	
17	8 points	
18	12 points	
19	2 points	

Question #	Max. points	Student score
20	2 points	
21	4 points	
22a	2 points	
22b	1 point	
subtotal	/ 40	
Ratios, Proportions, and Percent		
23	4 points	
24a	1 point	
24b	2 points	
24c	1 point	
24d	1 point	
25a	1 point	
25b	2 points	
26	Proportion: 1 point Solution: 2 points	
27	2 points	
28	2 points	
29	2 points	
30	2 points	
31	2 points	
32	2 points	
33	2 points	
subtotal	/ 29	
SUBTOTAL FOR THE FIRST FOUR SECTIONS:	**/120**	

Question #	Max. points	Student score
Geometry		
34a	2 points	
34b	2 points	
35	3 points	
36	2 points	
37a	2 points	
37b	2 points	
37c	1 point	
38	2 points	
39a	1 points	
39b	3 points	
40	2 points	
41	2 points	
42	3 points	
43a	2 points	
43b	2 points	
44	3 points	
45a	2 points	
45b	1 point	
46a	1 point	
46b	1 point	
47a	1 point	
47b	1 point	
subtotal		/ 41

Question #	Max. points	Student score
Probability		
48	3 points	
49a	2 points	
49b	1 point	
49c	1 point	
49d	1 point	
50	3 points	
51	3 points	
subtotal		/14
Statistics		
52	2 points	
53a	1 point	
53b	2 points	
53c	2 points	
54	2 points	
55a	1 point	
55b	1 point	
55c	1 point	
subtotal		/12
SUBTOTAL FOR THE LAST THREE SECTIONS:		**/67**
TOTAL		**/187**

End-of-Year Test — Grade 7
International Version

Integers

A calculator is not allowed for the problems in this section.

1. Give a real-life situation for the sum −15 + 10.

2. Give a real-life situation for the product 4 · (−2).

3. Represent the following operations on the number line.

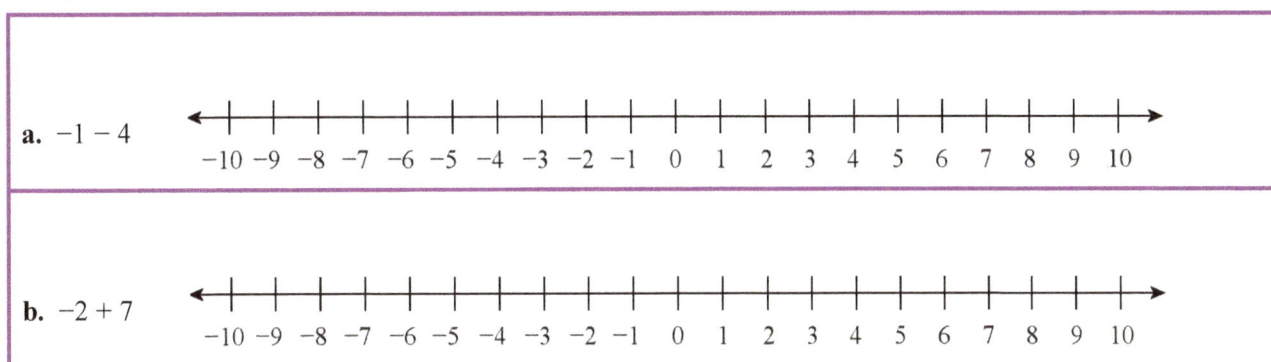

a. −1 − 4

$$\begin{array}{c} \xleftarrow{\hspace{1cm}} \quad | \quad | \quad | \quad | \quad | \quad | \quad | \quad | \quad | \quad | \quad | \quad | \quad | \quad | \quad | \quad | \quad | \quad | \quad | \quad | \quad | \xrightarrow{\hspace{1cm}} \\ \text{−10 −9 −8 −7 −6 −5 −4 −3 −2 −1 \ 0 \ 1 \ 2 \ 3 \ 4 \ 5 \ 6 \ 7 \ 8 \ 9 \ 10} \end{array}$$

b. −2 + 7

$$\begin{array}{c} \xleftarrow{\hspace{1cm}} \quad | \quad | \quad | \quad | \quad | \quad | \quad | \quad | \quad | \quad | \quad | \quad | \quad | \quad | \quad | \quad | \quad | \quad | \quad | \quad | \quad | \xrightarrow{\hspace{1cm}} \\ \text{−10 −9 −8 −7 −6 −5 −4 −3 −2 −1 \ 0 \ 1 \ 2 \ 3 \ 4 \ 5 \ 6 \ 7 \ 8 \ 9 \ 10} \end{array}$$

4. Solve.

a. −13 + (−45) + 60 = _____	**b.** −8 − (−7) = _____	**c.** 2 − (−17) + 6 = _____
d. −3 · (−8) = _____	**e.** 48 ÷ (−4) = _____	**f.** (−2) · 3 · (−2) = _____
g. 2 · (−22) − 5 · 4	**h.** $-16 + \dfrac{36}{4-13}$	**i.** $\dfrac{(-4)}{8} + 2 \cdot 5$

5. The expression | 20 − 31 | gives us the distance between the numbers 20 and 31.
   Write a similar expression for the distance between −5 and −15 and simplify it.

6. Numbers *a* and *b* are integers that are 15 units apart. Number *a* is positive and *b* is negative.
   Both have an absolute value less than 9. What can their values be?

7. Divide. Give your answer as a fraction or mixed number in lowest terms.

a. 1 ÷ (−8)	**b.** −4 ÷ 16	**c.** −21 ÷ (−5)

Rational Numbers

A calculator is not allowed for the problems in this section.

8. Write the decimals as fractions.

a. 0.1748	b. −0.00483	c. 2.043928

9. Write the fractions as decimals.

a. $-\dfrac{28}{10\,000}$	b. $\dfrac{2493}{100}$	c. $7\dfrac{1338}{100\,000}$

10. Convert to decimals. If you find a repeating pattern, give the repeating part. If you don't, round your answer to five decimals.

a. $\dfrac{7}{13}$	b. $1\dfrac{9}{11}$

11. Give a real-life context for each multiplication or division. Then solve.

a. $1.2 \cdot 25$

b. $(3/5) \div 4$

12. Calculate. For problems with fractions, give your answer in lowest terms, and as a mixed number if applicable.

a. $-\dfrac{2}{7} \cdot \left(-3\dfrac{5}{8}\right)$	**b.** $27.5 \div 0.6$
c. $-0.7 \cdot 1.1 \cdot (-0.001)$	**d.** $(-0.12)^2$
e. $\dfrac{\dfrac{3}{4}}{\dfrac{5}{12}}$	**f.** $\dfrac{5\,\frac{1}{2}}{-\dfrac{7}{8}}$

13. Calculate. You may give your answer as a decimal or a fraction.

a. $-\dfrac{1}{6} \cdot 1.2$	**b.** $-\dfrac{2}{5} \div (-0.1)$

14. Calculate. You may give your answer as a decimal or a fraction.

a. $-\dfrac{1}{2} + \dfrac{5}{2} \cdot (-0.8)$	**b.** $\dfrac{5}{8} \cdot 0.4 \cdot \left(-\dfrac{2}{3}\right) - 1.25$

Algebra

A calculator is not allowed for the problems in this section.

15. Simplify the expressions.

a. $7s + 2 + 8s - 12$	**b.** $x \cdot 5 \cdot x \cdot x \cdot x$	**c.** $3(a + b - 2)$
d. $0.02x + x$	**e.** $\dfrac{1}{3}(6w - 12)$	**f.** $-1.3a + 0.5 - 2.6a$

16. Factor the expressions (write them as multiplications).

a. $7x + 14$ =	**b.** $15 - 5y$ =	**c.** $21a + 24b - 9$ =

17. Solve the equations.

a. $\quad 2x - 7 = -6$	**b.** $\quad 2 - 9 = -z + 4$
c. $\quad 120 = \dfrac{c}{-10}$	**d.** $\quad 2\left(x + \dfrac{1}{2}\right) = -15$

64

18. Solve.

a. $\frac{2}{3}x = 266$	b. $x + 1\frac{1}{2} = \frac{3}{8}$
c. $-5y + 9y - 2 + y = 10$	d. $2(x + 7) - 3x = -36$
e. $\frac{y + 6}{-2} = -10$	f. $\frac{w}{2} - 3 = 0.8$

19. Chris can run at a constant speed of 12 km/h. How long will it take him to run from his home to the park, a distance of 0.8 km?

Remember to check that your answer is reasonable.

20. This shape consists of a rectangle and a regular hexagon. Its perimeter is 72 units and the longer side of the rectangle is 15 units. Write an equation to solve for the side of the hexagon, and solve it.

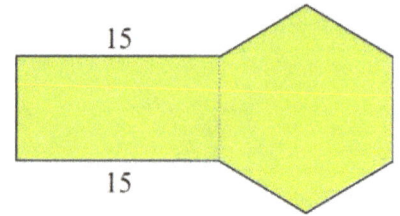

21. Solve the inequalities and plot their solution sets on a number line. Write appropriate multiples of ten under the bolded tick marks (for example, 30, 40, and 50).

a. $\quad 3x - 7 \; < \; 83$	**b.** $\quad -2x - 16.3 \; > \; 10.5$

22. You need to buy canning jars. They cost $19 per box, and you only have $170 to spend. You also have a coupon for a $25 discount on your total. How many boxes can you buy at most?

a. Write an inequality for the problem and solve it.

b. Describe the solution of the inequality in words.

Ratios, Proportions, and Percent

You may use a basic calculator for all the problems in this section.

23. Write a unit rate as a complex fraction and simplify it. Be sure to include the units.

a. Lily paid $15 for 3/8 kg of nuts.

b. Ryan walked 4 ½ kilometres in 3/4 of an hour.

24. The graph below shows the distance covered by a moped advancing at a constant speed.

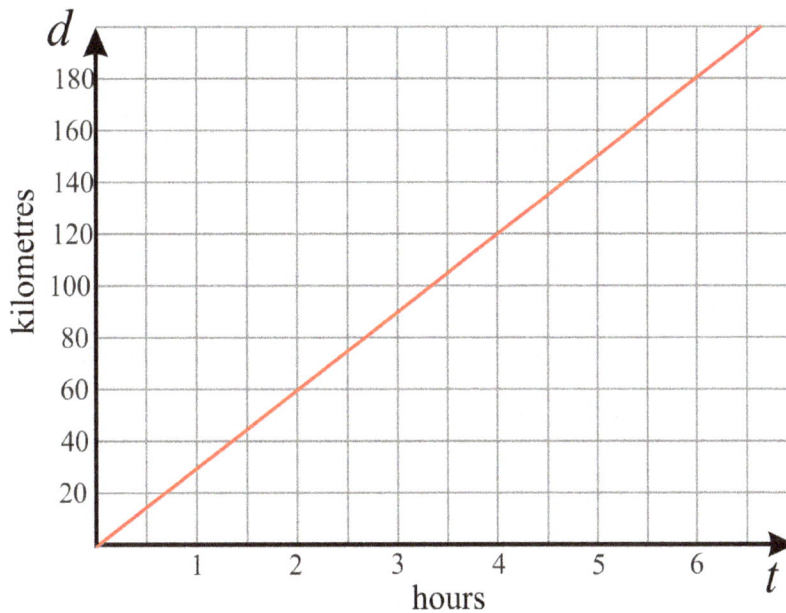

a. What is the speed of the moped?

b. Plot the point that corresponds to the time t = 4 hours.
What does that point signify in this context?

c. Write an equation relating the quantities d and t.

d. Plot the point that corresponds to the unit rate in this situation.

25. A Toyota Prius is able to go 904 km on 45.0 litres of petrol (highway driving).
A Honda Accord can travel 990 km on 65.0 litres of petrol (highway driving).
(Source: Fueleconomy.gov)

   a. Which car has better fuel efficiency?

   b. Calculate the difference in costs if you drive a distance of 480 km with each car,
   if petrol costs $1.482 per litre.

26. Write a proportion for the following problem and solve it.

   600 ml of oil weighs 554 g.
   How much would 5 litres of oil weigh?   ─────────── = ───────────

27. A farmer sells potatoes in sacks of various weights. The table shows the price per weight.

Weight	5 kg	10 kg	15 kg	20 kg	30 kg	50 kg
Price	$9	$18	$26	$32	$46	$70

   a. Are these two quantities in proportion?

   Explain how you can tell that.

   b. If so, write an equation relating the two and state the constant of proportionality.

28. Sally deposits $2500 at an 8% interest rate for 3 years (simple interest, not compound). How much can she withdraw at the end of that period?

29. A kitchen gadget was discounted by 18%, and then a 5.5% sales tax was added. The final price is $51.82. What was the original price before the discount and sales tax?

30. A ticket to a fair initially costs $20. The price is increased by 15%. Then, the price is decreased by 25% (from the already increased price). What is the final price of the ticket?

31. Margaret was shopping for a mattress, and found a store which had a 20% off sale going on. On top of that, the seller gave Margaret a further $30 discount, so she only paid $144. What was the price of the mattress originally?

32. Alex measured the rainfall on his property to be 10.5 cm in June, which he calculated to be a 35% increase compared to the previous month. How much had it rained in May?

33. In December, Sarah's website had 72 000 visitors. In December of the previous year it had 51 500 visitors.

   a. Find the percentage of increase to the nearest tenth of a percent in the number of visitors her website had for that year.

   b. If the number of visitors continues to grow at the same rate, about how many visitors (to the nearest thousand) will her site have in December of the following year?

Geometry

You may use a basic calculator for all the problems in this section.

34. The rectangle you see below is Jayden's room, drawn at the scale of 1:45.

a. Calculate the area of Jayden's room in reality, in square metres.
Hint: measure the dimensions of the rectangle in centimetres.

Scale 1:45

b. Do another scale drawing of Jayden's room, at a scale of 1:60.

35. A room measures 4 cm by 3 cm in a house plan with a scale of 1 cm : 0.8 m.
Calculate the actual dimensions of the room.

36. A square with 15-cm sides is enlarged with a scale factor of 4/3. What is the area of the resulting square?

37. A circle has a diameter of 16.0 cm.

   a. Calculate its area (to the nearest square centimetre) and circumference (to the nearest tenth of a centimetre).

   b. That circle is shrunk with a scale factor of 0.8. Calculate the area and the circumference of this smaller circle.

   c. What percentage is the area of the smaller circle of the area of the original?

38. Draw a triangle with sides 8 cm, 11 cm, and 14.5 cm using a compass and a ruler.

39. A triangle has angles that measure 36°, 90°, and 54°, and a side of 8 cm.

   a. Does the information given determine a unique triangle?

   b. If so, draw the triangle. If not, draw at least two different triangles that fit the description.

40. Three angles form a complete circle. Their measures are in the ratio of 5:2:8.
   Find the measure of those angles by writing an equation for the problem,
   and solve it.

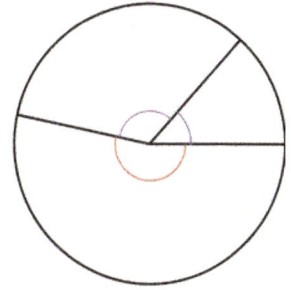

41. **a.** Write an equation for the measure of angle x,
   and solve it.

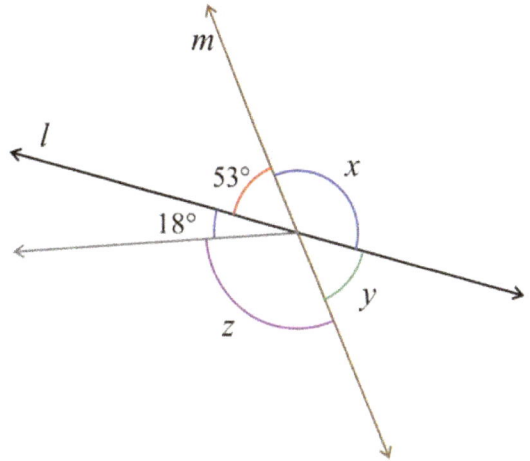

   b. Write an equation for the measure of angle z,
   and solve it.

42. Describe the cross sections formed by the intersection of the plane and the solid.

a.	b.	c.
The cross section is	The cross section is	The cross section is
_____.	_____.	_____.

73

43. **a.** Calculate the volume enclosed by
the roof (the top part).

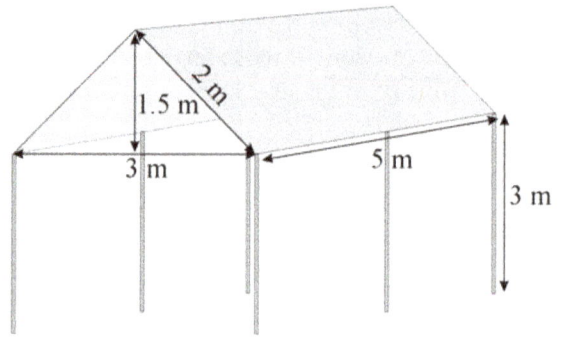

b. Calculate the total volume enclosed by the canopy.

44. Here you see a cube (with 2-unit edges) and a right triangular prism
that sits on top of that cube. The bottom face of the triangular prism
covers exactly half of the top face of the cube. Find the surface area
of this compound shape.

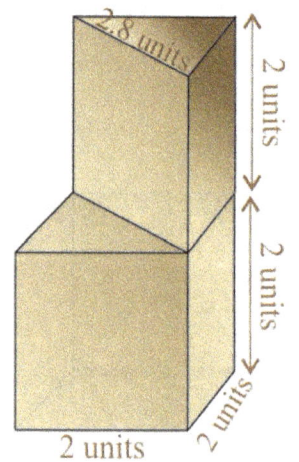

45. Two identical trapeziums are placed inside a 15 cm by 15 cm square.

   a. Calculate their area.

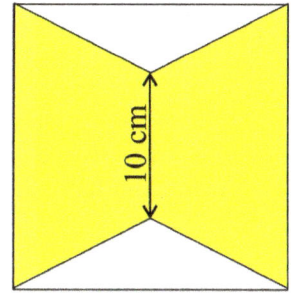

10 cm

   b. What percentage of the square do the trapeziums cover?

46. **a.** *Find the volume of the cylindrical part of the juicer, if its bottom diameter is 12 cm and its height is 4.5 cm.

   b. *Convert the volume to millilitres and to litres, considering that 1 ml = 1 cm^3.

47. **a.** How many cubic centimetres are in one cubic metre?

   b. Each edge of a cube measures 0.89 m. Calculate the volume of the cube in cubic centimetres.

Probability

You may use a basic calculator for all the problems in this section.

48. You randomly pick one marble from these marbles.
    Find the probabilities:

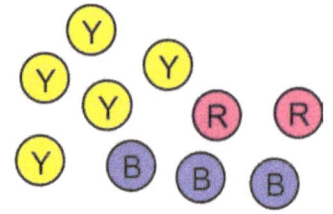

    a. P(not red)

    b. P(blue or red)

    c. P(green)

49. A cafeteria offers a main dish with chicken or beef. The customer then chooses a portion of rice, pasta, or potatoes, and a side dish of green salad, green beans, steamed cabbage, or coleslaw.

    a. Draw a tree diagram or make a list of all the possible meal combinations.

    A customer chooses the parts of the meal randomly. Find the probabilities:

    b. P(beef, rice, coleslaw)

    c. P(no coleslaw nor steamed cabbage)

    d. P(chicken, green salad)

50. John and Jim rolled a die 1000 times. The bar graph shows their results. Based on the results, which of the following conclusions, if any, are valid?

(a) This die is unfair.

(b) On this die, you will always get more 1s than 6s.

(c) Next time you roll, you will not get a 4.

A Thousand Dice Rolls

(bar graph: x-axis "Number Rolled" from 1 to 6, y-axis "Frequency" from 0 to 200)

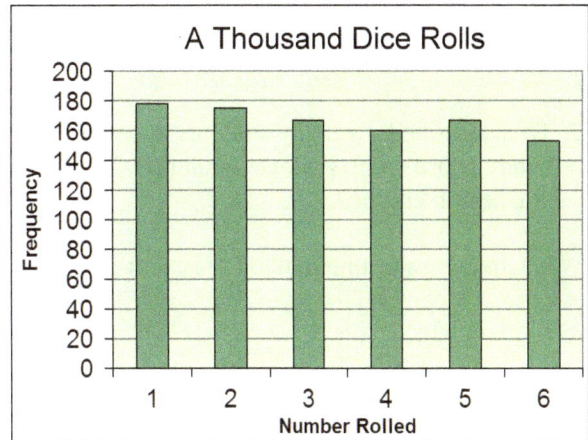

51. Let's assume that when a child is born, the probability that it is a boy is 1/2 and also 1/2 for a girl. One year, there were 10 births in a small community, and nine of them were girls. Explain how you could use coin tosses to simulate the situation, and to find the (approximate) probability that out of 10 births, exactly nine are girls. (You do not have to actually perform the simulation—just explain how it would be done.)

Statistics

You may use a basic calculator for all the problems in this section.

52. To determine how many students in her college use a particular Internet search engine, Cindy chose some students randomly from her class, and asked them whether they used that search engine.

    Is Cindy's sampling method biased or unbiased?

    Explain why.

53. Four people are running for mayor in a town of about 20 000 people. Three polls were conducted, each time asking 150 people who they would vote for. The table shows the results.

	Clark	Taylor	Thomas	Wright	Totals
Poll 1	58	19	61	12	150
Poll 2	68	17	56	9	150
Poll 3	65	22	53	10	150

    a. Based on the polls, predict the winner of the election.

    b. Assuming there will be 8500 voters in the actual election, estimate to the nearest hundred votes how many votes Thomas will get.

    c. Gauge how much off your estimate might be.

54. Gabriel randomly surveyed some households in a small community to determine how many of them support building a new highway near the community. Here are the results:

Opinion	Number
Support the highway	45
Do not support it	57
No opinion	18

If the community contains a total of 2120 households, predict how many of them would support building the highway.

55. A farmer had chickens (of the same breed) in two locations on his farm (pastures 1 and 2). He wanted to compare the weights of these two flocks of chickens, so he weighed every chicken in each flock. Here is a side-by-side boxplot showing the results.

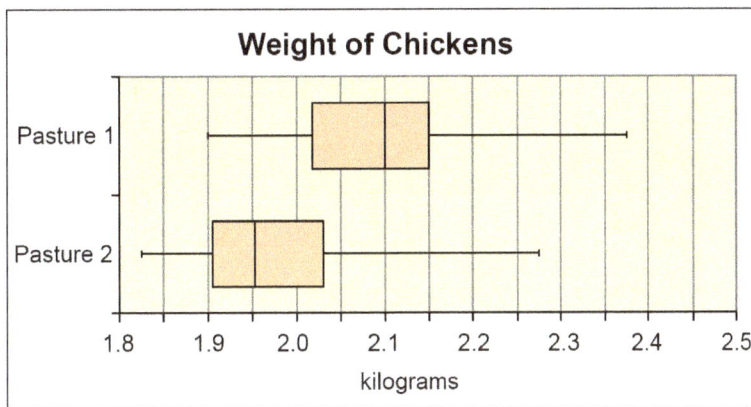

Weight of Chickens

a. Just looking at the two distributions, do chickens in either flock appear to weigh more? Explain how you know.

b. Do chickens in either flock appear to vary more in weight? Explain.

c. Does one of the flocks have *significantly* heavier chickens than the other?

If so, which one?

Justify your reasoning.

USING THE CUMULATIVE REVISIONS

The cumulative revisions include problems from various chapters of the Math Mammoth complete curriculum, up to the chapter named in the revision. For example, a cumulative revision for chapters 1-6 may include problems from chapters 1, 2, 3, 4, 5, and 6. The revision for chapters 1-6 can be used any time after the student has gone through chapter 6.

These revisions provide additional practice and revision. The teacher should decide when and if they are used—the student does not necessarily have to complete all the revisions

I recommend using at least 2 or 3 of these revisions during the school year. The teacher can also use the revisions as diagnostic tests to find out what areas and topics the student has trouble with.

Math Mammoth complete curriculum also comes with an easy worksheet maker, which is the perfect tool to make lots of problems of a specific type, especially when it comes to calculation skills.

You can access the worksheet maker online at

https://www.mathmammoth.com/private/Make_extra_worksheets_grade7.htm

Please note that while the worksheet maker covers a lot of topics, it does not include worksheets for all the topics in the curriculum. Most notably, it does not make worksheets for word problems. However, most people find it to be a very helpful addition to the curriculum.

Cumulative Revision, Grade 7, Chapters 1-2

1. Rewrite each expression using a fraction line, then simplify.

a. $7 \div 8 \cdot 4$	**b.** $5 \cdot 2 \div 10 + 1$	**c.** $(10 + 3) \div (8 - 1)$

2. Evaluate the expressions. (Give your answer as a fraction or mixed number, not as a decimal.)

a. $\dfrac{x+2}{x-2}$, when $x = 21$	**b.** $3s^2 - 2t^2$, when $s = 10$ and $t = -3$

3. Name the property of arithmetic illustrated by the equation $(5x)y = 5(xy)$.

4. There are two broomsticks, one wooden and one metal.

   a. Choose two variables to denote the lengths of the two broomsticks.

   Let _____ be the length of the wooden broomstick.

   Let _____ be the length of the metal one.

   b. Write an equation that matches the sentence "The wooden broomstick is 20 cm longer than the metal one."

5. **a.** Circle the equation that matches the situation.

   Let p be the normal price of one sun hat in a clothing store.
   The store owner decides to discount them by $5 each.
   A customer buys three sun hats. The total cost is $16.80.

   $3(p - \$5) = \$16.80$ $p - \$5 = 3 \cdot \$16.80$

   $3p - \$5 = \$16.80$ $3(p - 0.5) = \$16.80$

   b. How much would one sun hat have cost before the discount?
   Solve this problem using any strategy. You don't have
   to use the equation.

6. Simplify the expressions.

a. $6p + 2 + 5p - 1$	**b.** $6p \cdot p \cdot 7$	**c.** $f \cdot 2f \cdot 2f \cdot f \cdot 3$

7. Simplify.

a. $|-71|$	**b.** $-|-2|$	**c.** $|-9 + 5|$	**d.** $-(-84)$
e. $|-9| + |-5|$		**f.** $|-9| - |5|$	

8. Write an inequality. Use negative integers where appropriate.

   a. This hill is at least 60 m high.

   b. Liz owes more than $120.

   c. A maximum of 8 items per customer.

   d. The ride is only for children that are up to 120 cm tall.

9. Fill in the missing numbers.

a. $5 - $ _____ $= {}^-2$	**c.** $2 + $ _____ $= {}^-4$	**e.** $^-30 + $ _____ $= {}^-40$	**g.** $^-51 + $ _____ $= 0$
b. $^-1 - $ _____ $= {}^-19$	**d.** $4 - $ _____ $= 5$	**f.** $0 - $ _____ $= {}^-49$	**h.** $^-9 + $ _____ $= {}^-7$

10. Answer the questions about the pattern.

Step   1   2   3   4   5

a. Draw steps 4 and 5.

b. How do you see this pattern growing?

c. How many flowers will there be in step 39?

d. What about in step n?

Cumulative Revision, Grade 7, Chapters 1-3

1. Find the root(s) of the equation $x^2 - x - 20 = 0$ in the set $\{-5, -4, -3, 3, 4, 5\}$.

2. Is division commutative? Explain your reasoning.

3. Solve.

a. $\quad \dfrac{1}{7}x \;=\; -15$	**b.** $\quad 3 - (-8) \;=\; \dfrac{x}{-12}$
c. $\quad 7 - x \;=\; -3$	**d.** $\quad 5 \cdot (-8) \;=\; -6x - 4x$

4. A rectangle has sides 15 units and $x + 3$ units long. Write an expression for the area of a rectangle and simplify it.

5. **a.** Write an expression using mathematical symbols: "the quantity $2x$ minus 1, squared."

   b. Evaluate the expression when x is -4.

6. Amanda swims 1 kilometre in 35 minutes.
   Find her average speed in kilometres per hour.

7. Consider the four expressions $67 + 28$, $(-67) + (-28)$, $(-67) + 28$, and $67 + (-28)$. Write these expressions in order from the one with **least** value to the one with **greatest** value.

8. Add.

a. $(-3) + (-6) + 5 + 1 =$ _____	**b.** $14 + (-14) + (-31) + 11 + (-6) =$ _____

9. Divide and simplify if possible.

a. $60 \div (-5)$	**b.** $-33 \div 15$	**c.** $-2 \div (-9)$

10. Jerry's yard is a rectangle with one 40-m side and a total area of 1000 m^2.
    How long is the other side? Write an equation with an unknown and solve it.

Cumulative Revision, Grade 7, Chapters 1-4

1. Are the expressions equal, no matter what values n and m have? If so, you don't need to do anything else. If not, provide a counterexample: specific values of n and m that show the expressions do NOT have the same value.

a. $(-n - 1) - m$ $-n - (1 - m)$	**b.** $\dfrac{x - y}{2}$ $\dfrac{x}{2} - \dfrac{y}{2}$

2. **a.** Sketch a rectangle with sides $3s$ and $8s$ long.

   b. What is its area?

   c. What is its perimeter?

3. Simplify the expressions.

a. $23r - 8r + 7r + 5$	**b.** $9p^2 + 8 - 3p^2$	**c.** $6y \cdot y \cdot 7y$

4. In each exercise (a) and (b), write a single expression to match the two number line jumps, and solve.

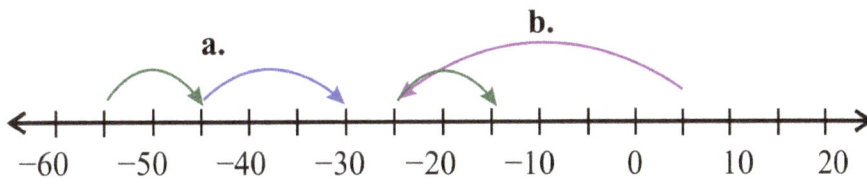

a.

b.

5. Ashley has a bunch of 20-cent coins. The total value of her coins is 860 cents.
   How many 20-cent coins does Ashley have?
   Write an equation and solve it.

6. Factor these expressions (write them as products).

a. $100x + 60 =$	**b.** $24s - 4t - 8 =$

7. Calculate.

a. $-2 \cdot (-10) \cdot 5 \cdot (-20) =$	**b.** $(-2)^5 =$	**c.** $(-2)^3 \cdot (-3) =$

8. In a two-team card game, Team 1 had -49 points and
   Team 2 was 154 points ahead of the first team.
   How many points did Team 2 have?

9. Which expression(s) give us the distance between x and 8? (We don't know if x is negative, positive, or zero.)

a. $x - 8$	**b.** $|x + 8|$	**c.** $x - (-8)$	**d.** $|8 - x|$	**e.** $|x - 8|$	**f.** $|x - (-8)|$

10. Solve using both decimal division and fraction division.

$1.3 \div 0.3$	**Decimal division:**	**Fraction division:**
	$\overline{)}$	

11. Give a real-life context for the division in exercise #3.

12. Solve *without* a calculator.

a. 30% of $400	**b.** $\dfrac{2}{3} \cdot 6.9$ km	**c.** $-0.08 \cdot \dfrac{1}{10} \cdot 5$

88

Cumulative Revision, Grade 7, Chapters 1-5

1. **a.** Mark two points on the number line that are 0.4 units from point A.

   b. Mark two points on the number line that are the same distance apart as points A and B, and are both positive.

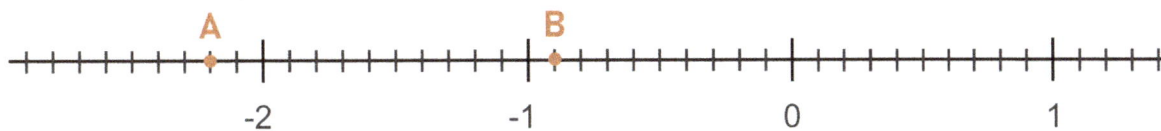

2. Multiply.

a. $(-7) \cdot 2 \cdot (-2)$	**b.** $10 \cdot (-4) \cdot 7$	**c.** $2 \cdot (-5) \cdot (-2) \cdot (-5)$

3. Solve, using the correct order of operations.

a. $-22 + 3 \cdot (-7) =$ _____	**b.** $6 \cdot 5 - 9 \cdot 5 =$ _____	**c.** $-19 - 3 \cdot (-5) =$ _____

4. Below you see listed the minimum daily temperatures for one week. Calculate their average.

   $-8°C, -11°C, 2°C, 0°C, -3°C, -5°C, -1°C$

5. Let $x = -3$ and $y = 7$. Evaluate the expressions.

a. $x + y$	**b.** $x - y$	**c.** $y - x$	**d.** $|y - x|$

6. Karen has 10 1/2 cups of flour left and she is planning to make biscuits to sell. How many batches of biscuits can she make with that, if one batch takes 2 1/4 cups of flour, *and* she needs to leave at least one cup of flour for something else?

   Also, how much flour will she have left after making the biscuits?

7. Multiply mentally.

a. $0.3 \cdot 2.5$	**b.** $-0.002 \cdot 0.008$	**c.** $-0.9 \cdot 50$
d. 0.8^2	**e.** $-4 \cdot 0.05 \cdot (-20)$	**f.** $(-0.3)^2$

8. Write the fractions as decimals.

a. $-\dfrac{61}{100\ 000}$

b. $\dfrac{9\ 807\ 200}{1000}$

c. $\dfrac{55\ 191}{1\ 000\ 000}$

9. Alex commutes 12 km to work every day. One day, his average speed going to work was 60 km/h and coming back 50 km/h. How many minutes did it take Alex to commute that day?

10. Solve. Give your answers as decimals, rounded to three decimal digits.

a. $5.5(x-1) = 7.6$	**b.** $-10(y+1.7) = 20$	**c.** $7(x+0.5) - 6.2x = 2$

11. Solve.

a. $\dfrac{x}{8} - 2 = -9$	**b.** $-15 = \dfrac{s}{7} - 8$	**c.** $\dfrac{1}{6}(x-30) = -3$

12. Write in decimal form. Use long division, and calculate each answer to at least six decimal places. If you find a repeating pattern, give the repeating part. If you don't, round your answer to five decimals.

a. $2\frac{5}{24}$	b. $2.05 \div 7$	c. $5.6 \div 0.02$

13. Solve each inequality. Plot the solution set on the number line.

a. $-5x + 3 \ < \ 22$	b. $24 - 10x \ > \ 49$

14. Add.

a. $\frac{3}{5} + \left(-\frac{2}{3}\right)$	b. $-\frac{1}{2} + \frac{5}{6} + \left(-\frac{6}{9}\right)$

Cumulative Revision, Grade 7, Chapters 1-6

1. Simplify the expressions.

a. $-3z - 9 + 7z + 2t$	**b.** $6x \cdot x \cdot (-7x)$	**c.** $6s \cdot s \cdot 4t$

2. Write the unit rate as "cups of flour per recipe".

Crystal made the muffin recipe 1 1/2 times and used 2 1/4 cups of flour.

3. Solve. Check your solutions.

a. $\quad 0.2 - 0.6 = 0.4(6x - 10)$	**b.** $\quad 3(x - 6) - 2x = -9$
c. $\quad 8x = -\dfrac{3}{4}$	**d.** $\quad 1\dfrac{1}{6} + v = \dfrac{2}{9}$

4. The floors of a skyscraper are 3.35 m apart. The bottom floor, which is actually the basement, is located 2.7 m below the ground. Write an equation to find which floor is at an elevation of 87.75 m, and solve it.

5. Cynthia took 14 minutes to bicycle from her home to a dentist appointment (a distance of 2.8 km) and 10 minutes to bicycle from there back home. Calculate her average speed for the entire trip.

6. A contractor has quoted you a price of $40.50 per square metre for an asphalt driveway. The driveway will be 8 m long but you need to decide the exact width. You can afford to spend at most $1600. What is the maximum width for the driveway?
Write an inequality for the problem and solve it.

7. An aeroplane travels at a constant speed of 900 km/h.

   a. Write an equation relating the distance (d) it has travelled and the time (t) that has passed.

   b. Plot your equation. Notice that you need to scale the d-axis.

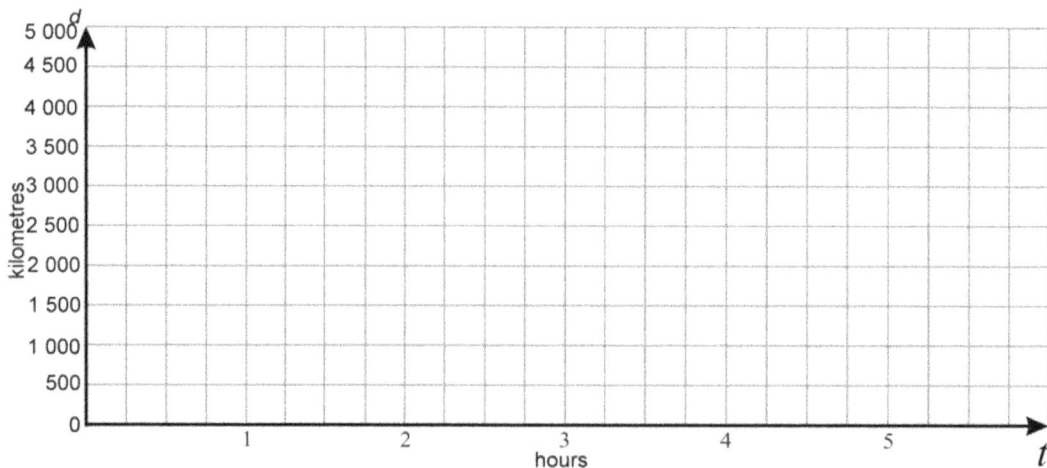

   c. How long will it take the aeroplane to travel 5000 km?

Cumulative Revision, Grade 7, Chapters 1-7

1. Solve each inequality and plot its solution set on the number line. Write appropriate multiples of ten under the bolded tick marks (for example, 30, 40, and 50).

a. $\quad 3y + 7 \;<\; 56$	**b.** $\quad -5 - 6z \;\geq\; 175$

2. Working at an art gallery, you are paid a base salary of $220 per week plus a $75 commission for every painting you sell. How many paintings would you need to sell in a week in order to earn at least $800? Write an inequality for this problem and solve it.

3. Solve (without a calculator). Use estimation to check that your answer is reasonable.

a. $10 - \dfrac{5}{6} \cdot 2.7$	**b.** $0.4 \div \left(\dfrac{2}{9} + \dfrac{1}{3} \right).$

4. Find the percentage of increase or decrease.

a. A flashlight that costs $9 is discounted so that now it costs $8.10. What is the discount percentage?	**b.** A chair used to cost $40, but now it costs $48. What is the percentage of increase?

5. The two triangles are similar. How long is the unknown side?

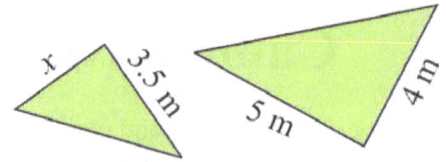

6. Redraw this rectangle at the scale 1:2.

Scale 1:5

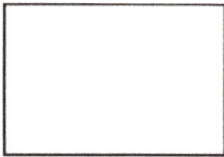

7. The table shows the number of workers and the time it takes them to paint a house.

Number of workers	1	2	3	4	5	6
Time (hours)	10	5	3.3	2.5	2	1.7

Are these two quantities in proportion?
How can you tell?

8. Solve.

a. $13 + (-37) + (-8) =$	**b.** $-94 - (-8) =$	**c.** $20 - 60 + 90 =$

9. Find the missing numbers.

a. $6 \cdot \underline{\hspace{1cm}} = -42$	**b.** $-72 \div \underline{\hspace{1cm}} = 8$	**c.** $\underline{\hspace{1cm}} \div (-12) = -4$

10. Find the value of the expressions when $x = -5$ and $y = 2$.

a. $1 - x^2$	b. $10xy$	c. $-3(x + y)$

11. Write an expression for the final price using a decimal for the percentage.

a. A bag of dog food: price p, discounted by 11%. New price = _____

b. Sunglasses: price s, price increased by 6%. New price = _____

12. Mariah bought a computer at a 25% discount. It also had a 4% sales tax added. The total was $514.02. What was the price of the computer before the discount and the sales tax? Write an equation for this situation and solve it.

13. The cost of a taxi fare is proportional to the distance driven. If a 5.6-km ride costs $28.56, then how much will an 8.8-km ride cost?

14. Matt got a really unreasonable answer for the problem below. Find what went wrong with his solution and correct it.

Jim can swim 30 laps in a pool in 26 minutes.
How many laps could he swim in 45 minutes?

Matt's Answer: He could swim 30 laps.

Solution:
$$\frac{30 \text{ laps}}{26 \text{ min}} = \frac{L}{45 \text{ min}}$$

$$26L = 30 \cdot 26$$

$$26L = 780$$

$$\frac{26L}{26} = \frac{780}{26}$$

$$L = 30$$

Cumulative Revision, Grade 7, Chapters 1-8

1. **a.** Sketch a rectangle with sides $5x$ and $6x$ long.

  b. What is its area?

  c. What is its perimeter?

2. A photo editing software was discounted by 2/5. The discounted price is $29.97.

  a. Find the original price using logical reasoning and/or a bar model.

  b. Choose a variable to represent the original price. Write an equation for the situation and solve it. Compare the solution steps of the equation to the way you solved the problem in (a).

3. Study the problem below and Jane's solution for it. She figured that exactly half of the books were to be sent to book stores.

A printing press printed 1500 copies of a book. 5/6 of those were printed as paperbacks and the rest were printed with hard covers. Now, 3/5 of the paperbacks need to be sent to various book stores. *How many books is that?*	Jane's calculation to solve this: $\dfrac{\overset{1}{\cancel{3}}}{\underset{1}{\cancel{5}}} \cdot \dfrac{\overset{1}{\cancel{5}}}{\underset{2}{\cancel{6}}} \cdot 1500 = \dfrac{1}{2} \cdot 1500 = 750$

  a. Is her answer correct?

  b. Is the way she calculated it correct?
    If not, correct it.

4. **a.** Draw a rectangle with an aspect ratio of 4:3 (width to height) so that the width is 6 cm.

   b. Then enlarge your rectangle using the scale factor 5/3. Draw the enlarged rectangle.

   c. What is the aspect ratio of the enlarged rectangle?

5. In an educational game, you earn 3 play coins for each 10 problems you complete.

   a. Give the unit rate as problems per coin.

   b. How many problems would you need to solve in order to earn 75 coins so you can purchase a phone for your virtual character in the game?

   c. In the same game, if you have solved 376 problems, how many coins have you earned?

6. A hexagonal prism is 12 cm high. Its base is a regular hexagon with a side of 2.5 cm and an area of 16.2 cm^2.

   a. Calculate the surface area of the prism.
   Hint: Sketch the prism to help you.

   b. Calculate the volume of the prism.

7. Which covers more of the wall, a square clock with 30-centimetre sides or a circular clock with 36-centimetre diameter?   How much more?

8. Mason got 16 points out of 21 in a quiz. What is his percentage score, to the nearest tenth of a percent?

9. Mark pays 22.5% of his income in taxes. If he earns $2350 in a month, find how much he has left after taxes.

10. A farmer gets paid $3 for a bushel of corn. A bushel is about 35.2 litres.

   a. How much does the farmer get for one litre of corn?

   b. Let P be the the amount of money the farmer gets and V be the amount of corn in litres. Write an equation relating the two variables.

   c. Plot your equation. Choose appropriate scaling for the P-axis.
   Hint: calculate how much the farmer gets for 100 litres and for 600 litres of corn.

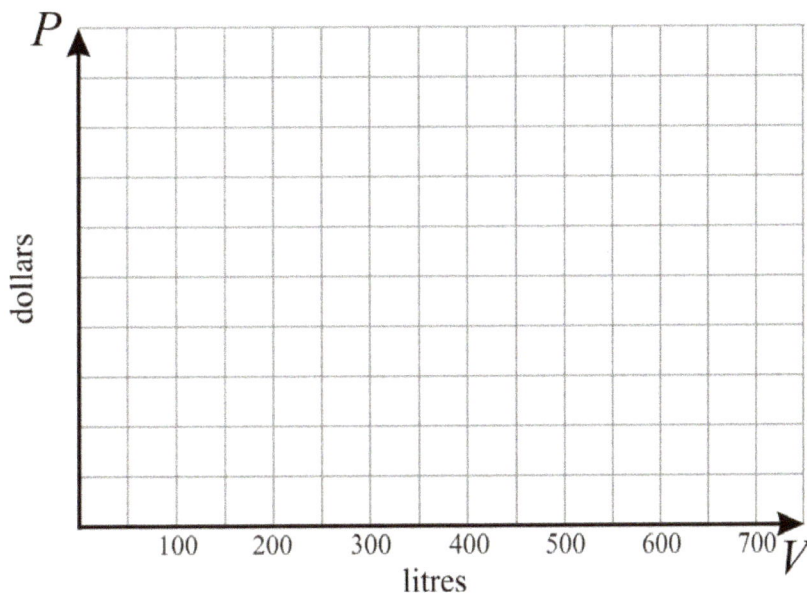

   d. Plot the point corresponding to 200 litres of corn.

   e. Plot the point corresponding to the farmer earning $50.

Cumulative Revision, Grade 7, Chapters 1-9

1. Jayden paid 1/7 of his salary in taxes, and then an additional $235 for a loan payment.
   Now he has $2180 left. How much was his salary?

   Write an equation for the situation and solve it.

2. Factor these expressions (write them as products).

a. $7x + 21$ = _____ (_____ + _____)	**b.** $24w - 16$ =
c. $-21t - 7$ =	**d.** $50a - 70b - 120 =$
e. $-55a + 30$ =	**f.** $-56y - 84 - 7x =$

3. Ava used 3 1/2 cans of paint to paint 2/3 of a room.

   a. Write the unit rate as a complex fraction and simplify it.

   b. Using paint at the same rate, how much more paint does Ava need to finish painting the room?

4. Two lines intersect at O, and additionally the figure has ray OA.
   Find the measure of angle AOB.

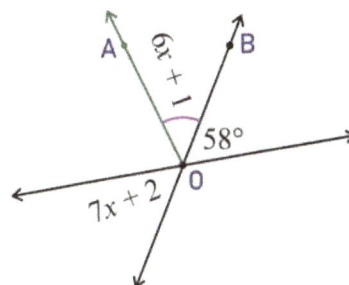

5. **a.** What is the area of a 1.2 metre by 0.8 metre rectangle in square centimetres?

    b. A triangle's base measures 1.5 m and its altitude is 0.75 m. What is its area in square centimetres?

6. Evaluate the expressions.

a. $100 - x^2$, when $x = -2$	**b.** $\dfrac{2w}{w + 3}$, when $w = -15$

7. Solve.

a. $x - \dfrac{5}{6} = 7\dfrac{1}{3}$	**b.** $5y = -\dfrac{4}{9}$	**c.** $\dfrac{1}{5}v - 2 = -7$

8. You choose the digits for a two-digit number randomly from the digits 2, 3, 5, and 7 (prime numbers). Each digit can be used twice; for example, it is possible to make 55.

    a. What is the probability of making a number between 31 and 40?

    b. What is the probability of making a number where both digits are the same?

9. The table below shows how long it takes for a car to travel a distance of 120 km at different speeds.

Speed (km/h)	120	100	80	60	40	20
Time (h)	1	1.2	1.5	2	3	6

a. Are these two quantities, speed and time, in proportion?

Explain how you can tell that.

b. If so, write an equation relating the two and state the constant of proportionality.

10. Three mirrors are set up in a form of an equilateral triangular prism. They will go inside a tube to make a kaleidoscope.

a. Calculate the volume of the triangular prism, to the nearest ten cubic centimetres.

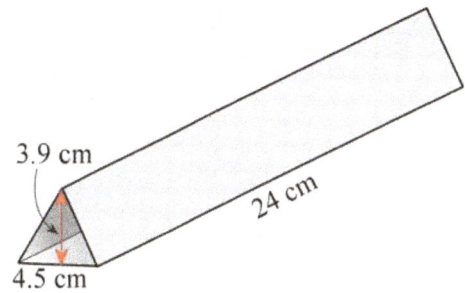

3.9 cm

24 cm

4.5 cm

b. Calculate its surface area, to the nearest ten square centimetres.

11. Draw an equilateral triangle using only a compass and a straightedge. Make the side length whatever you wish. If you draw a small one, you can draw it here. Or, you can use blank paper.

12. A triangle has 45° and 60° angles and a 5.6-cm side between those angles. Does the information given define a unique triangle? If it does, write yes, and draw the triangle.

   If not, prove that it doesn't by sketching at least two non-congruent (different-shaped) triangles that satisfy the given conditions.

13. Explain how we can use the pictures at the right to show that the area of a circle is A = ½C · r, where C is the circumference and r is the radius of the circle.

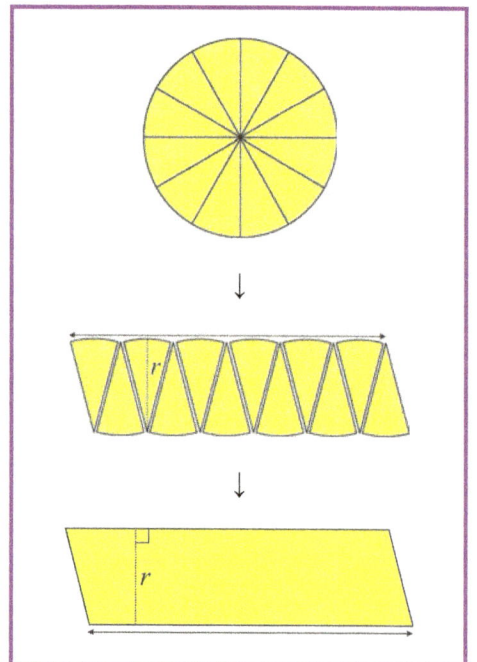

14. This shape consists of 3/4 of a circle and a square. The diameter of the circle is 4.2 cm. Calculate the area of the shape to the nearest square centimetre.

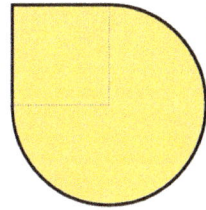

15. The price of a standing lamp increased from $19 to $22.50. What was the percent of increase?

16. Mason took a $1500 loan at 9.8% annual interest rate to purchase a computer. He paid it back 1 1/2 years later. Calculate the total amount Mason paid at that point.

17. Alison pays $65 per month for an Internet service and she can use 1200 gigabytes of bandwidth for that fee. For any bandwidth she uses in excess of that 1200 gigabytes, she will pay $10 per 50 gigabytes. She used 75% of her monthly bandwidth quota in 18 days in September. If she continues to use bandwidth at the same rate, how much extra will she pay at the end of the month?

18. The two figures are similar. Calculate the length of side marked with x.

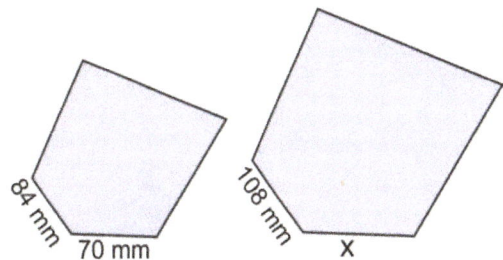

84 mm
70 mm
108 mm
X

Cumulative Revision, Grade 7, Chapters 1-10

You may use a basic calculator for all the problems in these worksheets.

1. The line graph shows the number of candles that a candle factory sold in years 2010 to 2015.
   Note that the scale is given in "thousand candles."

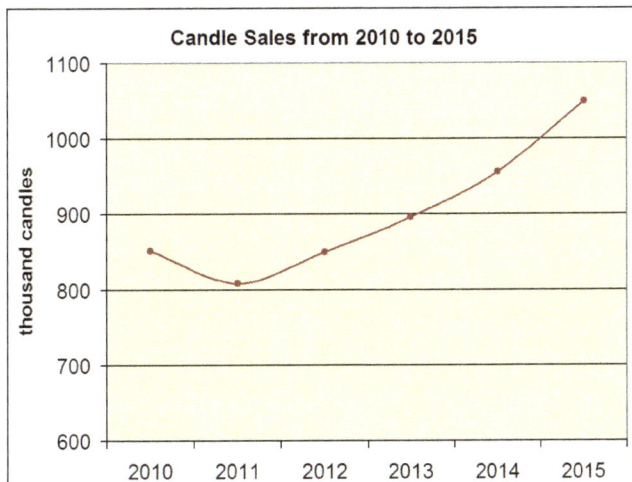

Candle Sales from 2010 to 2015

Estimating the amounts from the graph, calculate the approximate
percentage increase in the candle sales from 2010 to 2015.

2. If the edges of this cube measure 25 cm, how many such cubes do you
   need in order to have 1 cubic metre?

3. Solve the equations.

a. $\dfrac{1}{5}x + 12 = -14$	**b.** $\dfrac{w}{-0.2} = -0.4$
c. $7 = -4(x + 5) - 2x$	**d.** $\dfrac{x + 1}{4} = -2$

4. **a.** Redraw this shape at the scale of 1:16.

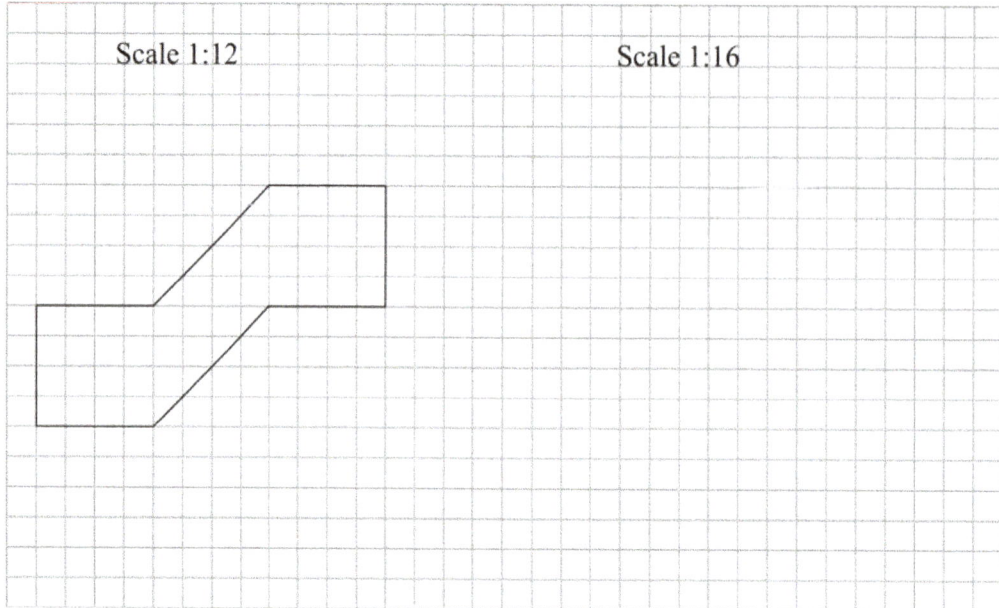

Scale 1:12   Scale 1:16

b. Find the area of the shape in reality, to the nearest ten square centimetres.
(If the page is printed at 100%, one unit in the grid measures 4 mm.)

5. Lines AC and BD meet at O.

   a. Write an equation for y (measure of $\angle COD$) and solve it.

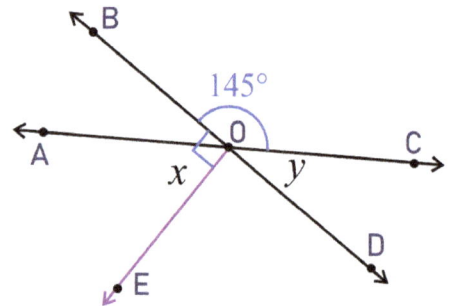

   b. Write an equation for x (measure of $\angle AOE$) and solve it.

6. A family gave the waiter a 15% tip on the $80 they were charged for a meal in a restaurant.
Plus they paid sales tax of 6.8% (on the $80). What was the total amount they spent?

7. **a.** What two-dimensional shape(s) are produced as a cross-section
   when this prism is cut with a vertical plane?

   b. What if it is cut with a horizontal plane?

8. **a.** Are these two typing rates equal: 60 words per 90 seconds and 135 words per 3 minutes?

   b. If not, calculate how many more words in 5 minutes you would type at the faster rate
   than at the slower rate.

9. The principal of a school wants to ask the students' parents whether some extra money should be used
   to purchase more books for the library, upgrade the computer systems, or to improve the sports facilities
   of the school. He randomly surveyed some parents who were attending a basketball game at the school.

   Is the principal's sampling method biased or unbiased?

   Explain why.

10. A bag contains 5 blue socks, 2 red socks, and 6 white socks.

    a. You draw one randomly. What is the probability that the sock is white?

    b. The first sock was indeed white! You draw another one randomly.
    What is the probability that this one is white?

    c. You draw two socks randomly. What is the probability of getting two socks of the same colour?

11. Researchers took a sample of 11 fruit drinks and 11 pop drinks to study the amount of sugar in these type of drinks. The dot plots show the amount of sugar in the sample drinks (250 ml portion of drink).

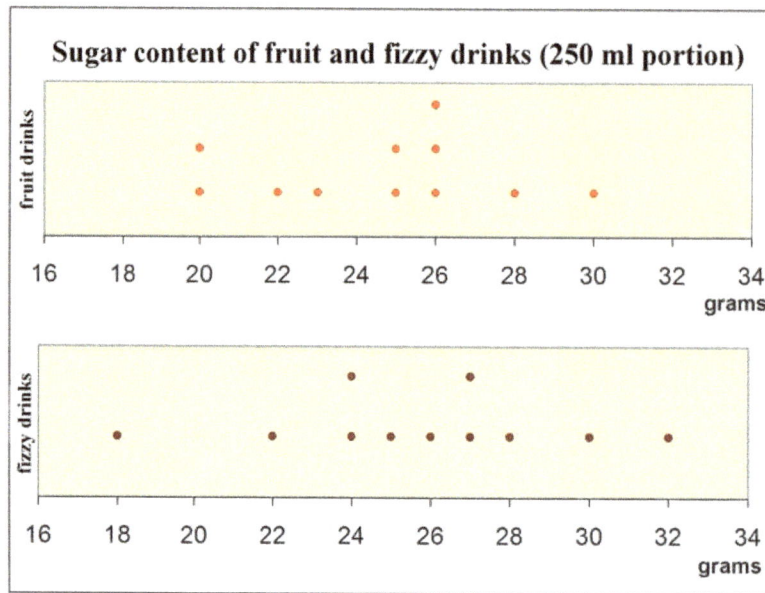

Sugar content of fruit and fizzy drinks (250 ml portion)

a. Based on the plots, which type of drink, if any, tends to contain more sugar?

Justify your answer.

b. Which type of drink, if any, has greater variability in the amount of sugar it has?

Justify your answer.

12. Let's say that 1/6 of the people in your region have hazel eyes. You randomly select five people. Design a simulation to find the following probabilities, and carry it out:

a. What is the probability that exactly two of those five people have hazel eyes?

b. What is the probability that none of those five people have hazel eyes?

13. Solve.

a. $\frac{2}{3}(x-8) = 30$	**b.** $6(y - \frac{2}{5}) = 24$	**c.** $21 = \frac{3}{4}(t-8)$

14. What is the area of the kitchen in reality, in square metres?

scale 1 : 250

15. A triangle has an 8-cm side, an 11-cm side, and a 40° angle. Does this information define a unique triangle? If yes, draw it. If not, sketch two non-congruent (differently shaped) triangles that fulfill the conditions.

16. Richard bought a dishwasher for $649 on credit with an 11.8% annual interest rate. How much interest (in dollars) will he pay per month?

17. Solve each proportion. Give your answers to the nearest hundredth.

a. $\dfrac{18}{47} = \dfrac{78}{s}$	b. $\dfrac{8.07}{11.4} = \dfrac{x}{605}$

18. Write in decimal form. Use long division, and calculate each answer to at least six decimal places. If you find a repeating pattern, give the repeating part. If you don't, round your answer to five decimals.

a. $\dfrac{42}{23}$	b. $6.05 \div 9$	c. $3.7 \div 0.11$

www.ingramcontent.com/pod-product-compliance
Lightning Source LLC
Chambersburg PA
CBHW080252200326
41519CB00023B/6961